Edward D. Cope

On the Contents of a Bone Cave in the Island of Anguilla

West Indies

Edward D. Cope

On the Contents of a Bone Cave in the Island of Anguilla
West Indies

ISBN/EAN: 9783337316044

Printed in Europe, USA, Canada, Australia, Japan

Cover: Foto ©Andreas Hilbeck / pixelio.de

More available books at **www.hansebooks.com**

SMITHSONIAN CONTRIBUTIONS TO KNOWLEDGE.

489

ON THE

CONTENTS OF A BONE CAVE

IN THE

ISLAND OF ANGUILLA

(*WEST INDIES*).

BY

EDWARD D. COPE.

WASHINGTON:
SMITHSONIAN INSTITUTION.
1883.

COMMISSION

TO WHICH THIS PAPER HAS BEEN REFERRED.

JOSEPH LEIDY.

ADVERTISEMENT.

THE following memoir describes the fossil vertebrates, shells, and indications of human occupation discovered during the excavation of a cave in the West Indian island of Anguilla.

The remains were first obtained in 1868, and brief notices of them have been made at various times since, but the publication of the full account was delayed in the hope that other objects might be added to the collection. The death of the gentleman who procured the specimens, and other causes having shown that no further exploration was practicable, the memoir was prepared and submitted to the Smithsonian Institution, for publication, in 1878.

The other works in progress prevented the publication until the present time, but the interval has been taken advantage of by the author to revise the manuscript and superintend the preparation of the plates.

The importance of the subject presented is shown by the following considerations :—

First—It is the first investigation of the life of the cave age in the West Indies.

Second—It gives the first reliable indication of the period of submergence, and hence of separation of the West Indian islands.

Third—It furnishes the first evidence as to the antiquity of man in the West Indies.

Fourth—It describes some very peculiar forms of animal life not previously known.

The illustrations have been made particularly full on account of the archæological interest attaching to those animals, which were probably the contemporaries of the earliest men of tropical America, and in order to avoid the necessity of any subsequent presentation of the same subject.

SPENCER F. BAIRD,
Secretary Smithsonian Institution.

WASHINGTON, April, 1883.

(iii)

CONTENTS OF A BONE CAVE.

In the year 1868, a quantity of cave earth, limestone fragments, and bone breccia was brought to the port of Philadelphia by a vessel and deposited on the lot of Henry Waters & Bro., manufacturers of phosphatic manures, on the Schuylkill. The material was imported for the purpose of ascertaining its value as a fertilizer, especially by the determination of its richness in calcium phosphate. It was obtained from a cave in the small Antillean island of Anguilla, which belongs to Denmark. Through the attention of Mr. Waters, I learned of the existence of fossil bones in the cargo, and proceeded to examine them. Remains of long bones lying irregularly in a rather hard but cavernous red cave deposit of limestone were found mingled with fragments of lighter limestone from the walls of the cave in irregular masses, the whole being penetrated and mixed with a yellow stalagmitic deposit of arragonite.

From a block of the breccia I dressed three molar teeth, two partially complete, and two much broken incisors, fragments of maxillary and pelvic bones, shafts of various long bones, and the distal extremity of a femur with a patella. These were the first evidences of the existence of the large rodent *Amblyrhiza inundata*, which was described in the proceedings of the American Philosophical Society for 1868. Other bones were found in other breccia masses, which I could not clearly refer to any other animal. With them occurred a shell of *Turbo pica*, Linn.

Having learned that Dr. E. van Rijgersma, colonial physician of the Danish Island of Saint Martins, was interested in all departments of the natural sciences, I wrote asking him to make an examination of the deposit in question, and to secure, if possible, all fossils discovered in excavating it. He accordingly very kindly went to Anguilla, and was rewarded by the possession of numerous additional teeth and bones of *Amblyrhiza*. Subsequent visits added two species of this genus, together with the bones of a species of ruminant near the genus *Capra*; bones of a probable rodent of smaller size, of two species of birds, of a lizard, and a shell chisel of human manufacture. These remains are described in the following pages.

Unfortunately no notes were taken as to the relations of the parts of the cave deposit, or whether any stratification is observable. We are, therefore, left to a consideration of the appearance of the fossils themselves in estimating their probable relative positions when found. The bones of the *Amblyrhiza invndata* first found were of an earthy-brown color. Those contained in the next sending were, some of them, of a brilliant vermilion color, and not encrusted. Others were of the usual earthy-brown color, and more or less encrusted; others of this lot were of a rusty red and not encrusted. The specimens of *Amblyrhiza quadrans* and *A. latidens*, subsequently discovered, are of a brownish-red, and are all encrusted with a thin calcareous layer. When this is removed their color is white. The bones of the ruminant and the shell-chisel are of the same superficial color, but were not encrusted.

As regards the mineral character of the bones, there is nothing to distinguish them from each other. They absorb moisture, so that broken surfaces adhere when brought into contact with the lips. A few specimens which do not display this character are not to be distinguished as to species from those that do. So there is no superficial character by which any distinction of age among these fossils can be made.

In the following pages the species are described in detail.

REPTILIA.

IGUANA, Linn.

The femur of a species of this genus was found at the time of the discovery of the *Amblyrhiza latidens*. In form and character it very closely resembles existing species of the genus.

The femur, which is that of the left side, has a gentle fore and aft sigmoid flexure. The head is laterally compressed, and is directed upwards and inwards. The trochanter is large, and forms a stout ala whose inferior extremity approaches and then sinks abruptly into the shaft. It is not curved, but its superior angle is thickened. It does not reach to the level of the angular margin which encircles the head. A low swelling marks the place of the little trochanter. The shaft is cylindric. The inferior extremity is transversely expanded, but there are no epicondyles. The condyles are but little contracted at the middle, but the external is much the larger. Its rotular face is entire, while that of the inner condyle has a shallow fossa on its posterior face. They are not separated by a popliteal notch, and have a very shallow trochlear groove anteriorly. There is a popliteal fossa. There is a slight epicondylar tuberosity on the superior external side of each condyle.

MEASUREMENTS.

		M.
Total length	.	.075
Long diameter of head . . .	.	.012
Diameter at apex of great trochanter .	.	.015
" of shaft	.	.007
" of condyles {fore and aft .	.	.006
{transverse .	.	.026

AVES.

Two bones of birds were discovered by Dr. van Rijgersma, one the coracoid of a larger and a humerus of a smaller species. The first is robust, but not shortened. Both its extremities are wanting, so that its determination is rendered difficult. It presents average proportions, not displaying the lateral expansion seen in the petrels nor the slender form of many waders. It resembles the corresponding portion of a *Grus*, and also the same part in some birds of prey. There is a well-defined inward-looking superior groove which is pierced by a large pneumatic foramen at a good distance from the margin. There is a corresponding superior foramen just below the scapular fossa. There is a rather large pneumatic fissure on the flat supero-internal face within the glenoid face.

The humerus above mentioned is that of a species of *Procellaria* in a restricted sense. The bird was of rather small size, and the characters of its humerus are well marked. Among these may be enumerated the prominent incurved posterior tuberosity of the head of the humerus, inclosing a deep fossa below it; the spur-shaped, anteriorly directed, acute epicondylar process of the inner side of the distal extremity. Within this spur is a strong coronoid fossa. The posterior border of the external condyle is produced well backwards.

MEASUREMENTS.

		M.
Length of humerus	.	.066
Width of proximal end	.	.014
" of shaft	.	.004
" of distal end	.	.008
Anteroposterior diameter of inner condyle	.	.005
" " of external condyle	.	.007
Length of epicondylar spur	.	.004

MAMMALIA.

As already observed, the mammalian bones are those of *Rodentia* and *Artiodactyla*. The determinable species are considered separately, but another of unknown reference will be first noticed.

This animal is represented by proximal and distal portions of opposite humeri, and both extremities of the femora, one of the latter being nearly entire. These appear to have belonged to a rodent of the size of an aguti, and perhaps allied to the group which that genus represents.

The head of the humerus is subround, and, with the proximal part of the shaft, a good deal incurved. The greater tuberosity is large, its rounded apex elevated above the head and rather prominent, but not incurved, in front. Its anterior margin soon joins the gently convex exterior margin of the shaft without irregularity, and the extero-posterior side of the shaft at this point is flat. The lesser tuberosity is quite small and separated from the outer by a wide bicipital groove.

The distal end of the humerus is not expanded, having small epicondyles. The trochlear face is strongly concave, ending in a deep olecranon, but no coronoid fossa. The bridge over the elongate arterial foramen is narrow.

The head of the femur is well bounded by a neck. The great trochanter rises as high as the head, and is deeply concave behind. The little trochanter is a prominent tuberosity on the posterior face of the bone. The shaft is slender and subcylindric. The condyles are somewhat compressed, and their surface is continuous with that of the patellar groove. The latter is longer than wide, moderately concave, with equal borders, which rise higher than the level of the shaft.

MEASUREMENTS.

	M.
Diameter of head of humerus .	.007
" of shaft of humerus . .	.008
" of epicondyles of humerus . .	.016
" of head of femur . .	.009
" at little trochanter of femur .	.012
" of shaft of femur . .	.008
" of condyles of femur . .	.017

AMBLYRHIZA, Cope.

Proceedings of the Philadelphia Academy, 1868, p. 313. Proceed. Amer. Philosoph. Society, 1869, p. 183. *Loxomylus*, Cope, Proceed. Amer. Philos. Soc., 1869, p. 186.

The characters of this genus are derived from the study of the remains of ten or more individuals of three species. These include portions of all parts of the skeleton with some important exceptions. There are wanting the superior, and parts of the inferior walls of the skull; the anterior foot, the distal parts of the pelvis, the calcaneum, and the ungual phalanges.

Skull and Teeth.—A fragment preserved includes the occipital condyle, with the mastoid and petrous bones. The condyloid foramen is pierced through the rather thin basi-occipital. The mastoid region is produced downwards and backwards, but the specimen being broken does not indicate its length. The petrous bone is in contact with the basi-occipital, closing the foramen lacerum posterius. Its inferior wall is not expanded bulla-like, but is flat and thick. The *meatus auditorius externus* is continued by the thick tympanic bone, in tubular form backwards and outwards. The apex of the petrous bone is truncate, and is deeply notched for the foramen. The premaxillary bones and the symphysis mandibuli are much produced and narrowed, and were probably inclosed in fur-bearing integument, as in the existing chinchillas. The mandibular rami are completely coössified. They are united at their lower borders, posterior to the divergence of their dental ridges, by the expansion of the stout rib of the inner face of each, which incloses the incisive alveolus. This ridge extends posteriorly very far: in the *A. latidens* to a point behind the last molar and even posterior to the base of the coronoid process. The latter is very short, and forms a flat process extending obliquely outwards from the ramus. The condyle of the mandible is narrow, and convex in the antero-posterior direction. There are some unattached hooked processes, which I suspect to be angles of the mandible.

The dental formula is i. $\frac{1}{1}$; c. $\frac{0}{0}$; m. $\frac{4}{4}$. The incisors have a moderately thick enamel layer, which is wrapped round the external angle a short distance. Their sculpture is not deeply cut. The molars are composed of vertical columns of dentine inclosed in and separated by a sheath of enamel. The columns are more or less transverse, and are neither confluent nor divided in any of the teeth. They number four in the superior teeth, excepting in the last molar where there are five. The entire tooth is inclosed in a thin layer of cement. The superior molars are curved bow-shaped, the convexity being directed forwards at the middle of their length. The enamel plates are then directed backwards on the grinding faces. The extremities of the roots are simple and contracted to an obtuse termination. The inferior molars differ in their form, being straight and directed obliquely forwards in the jaw. From this it results that their triturating surfaces are oblique to the axis of the teeth, while those of the superior molars are transverse to the axis of the middle portion of the shaft. There are but three columns in all of the inferior molars.

Vertebræ.—The atlas has widely expanded transverse processes whose base is about equally divided behind by the vertebral foramen. The vertebral canal soon issues on the inferior face, and again entering the transverse process near its middle, divides and issues in two foramina. One of these is on the superior aspect of the base of the transverse process, and the one opposite to it on the side of the neural canal. There is no hypapophysis nor *tuberculum atlantis*. The facets for the axis are directed obliquely inwards.

The axis is rather short and the neural spine long and directed upwards. Its section is triangular; the odontoid process is short and stout, with subcylindric section. Opposite articular extremity flat.

The lumbar centra are about as wide as long, and with nearly flat articular faces. The sacral vertebræ, although adherent in several individuals, preserve the suture in all of them. They are more elongate than the lumbars, and contract rapidly in diameter; the difference between the anterior and posterior dimensions of the first being considerable. The centra of all behind the first are much depressed, and the intervertebral foramina are large. The sacral diapophyses are greatly expanded, especially anteriorly; their iliac suture is a plane sloping inwards and upwards. The only caudal vertebra preserved is short and wide, has stout diapophyses, and no facets for chevron bones. As there is no trace of neurapophyses on this centrum, I infer that the tail is short.

Anterior Limb.—The scapula is represented by portions of those of several individuals. In these, the spine is well developed, rising rather abruptly from near the neck. The coracoid process is a short, stout hook. I have not found a clavicle among the bones.

The fore-limb is of smaller proportions than the posterior one. The greater tuberosity of the humerus is crest-like, extending along one side of the head, but not rising much above it. The lesser tuberosity is subround and very protuberant, inclosing a deep and wide bicipital groove, with the extremity of the greater tuberosity. The distal end of the humerus is moderately expanded in the transverse direction. The external epicondyle is protuberant; the internal wanting. The condyles have the ulnar and radial portions about equal, and the intertrochlear ridge is represented by an obscure angle. The external or ulnar keel is prominent. On the posterior face a rather short, median, concave, trochlear face represents the two anterior faces.

Both ulna and radius are rather slender; at the middle of their length the former is compressed, and the latter is depressed, both moderately. The olecranon is prominent and subcylindric. The head of the radius is a rather wide oval, directed downward and outward, while the coronoid process of the ulna is directed upward and inward. The fore-foot is unknown.

Posterior Limb.—The section of the neck of the ilium is an equilateral triangle, and is quite long; the crest is not preserved, but from the form of the sacrum may have had more extent than in many rodents. There is a strong tuberosity in the position of the anterior inferior spine. The *fossa ligamenti teris* and the *incisura acetabuli* are strongly developed, the latter not perforate. The posterior bones of the pelvis are unknown.

The femur has several marked peculiarities. One of these is the great development of the great trochanter, which is really an undiminished continuation of the shaft for some distance beyond the head. The head is relatively small, and is more than half a sphere; its ligamentous pit is distinct and isolated; the neck is abruptly contracted within the head, and projects at right angles to the shaft. The posterior base of the great trochanter is deeply excavated. The little trochanter is a prominent tuberosity. The condyles have a predominating inferior exposure. The patellar groove is wide and not elevated. There is a deep fossa in the external epicondylar region.

The head of the tibia is expanded laterally, and the upper part of the crest is replaced by a plane surface, as in the corresponding bone in man. The spine is prominent, a portion belonging to each of the articular faces, which are separated by a deep groove. The upper portion of the shaft is deeply grooved on the external face. The distal portion of the shaft is relatively slender. The distal extremity is expanded inwards. The astragalar facets are oblique, the external is larger than the internal, and they are well separated by an obtuse ridge. There are two processes on the internal border, which are separated by a deep tendinous groove, which is in most of the specimens bridged over by a lamina connecting the processes. The posterior of the two processes is the most elongate: it corresponds to a process of the astragalus, which extends backwards and inwards from the internal trochlear face. When extension of the foot is attempted, these processes come in contact, and prevent further movement. The amount of extension from the horizontal which this arrangement permits, is 45°. When at this point the processes constitute a support to the weight of the animal in addition to that furnished by the usual astragalar facets. The fibular plane is triangular, and has a posterior as well as exterior exposure.

The astragalus is quite depressed, and the convexity of its rotular surfaces is not great. The latter are of unequal size, the external having four times the extent of the internal, and they are well separated by the rotular groove, which is here formed by the meeting of planes of different angles. There is little indication of fibular surface on the external side of the astragalus, but the internal side of the inner rotula bears an oblique facet. This forms an angle of 91° with the rotula proper, which angle is directed upwards. Outwards and backwards from the base of this rotula extends the process already described. The head of the astragalus is short, narrow, and defined by a neck. Its navicular face is subround, slightly convex, and in the transverse plane. The head bears on its external side a facet for a bone, perhaps the cuboid, as in the aguti. On the inferior face the external calcaneal face is concave, and much larger than the internal, which is plane. They are completely separated by a groove which runs out on the postero-interior process already mentioned.

Of the hinder foot there is preserved a mass which includes the navicular, ecto- and mesocuneiform, and the second and third metatarsals. There is a loose fourth metatarsal, which presents a proximal face for the fifth. The metatarsals are stout, indicating a plantigrade foot. The navicular is not elongate, and has considerable horizontal extent. A part only of the proximal surface is occupied by the concave facet for the astragalus; the remaining portion extends outwards and backwards, and supports a couple of facets at its anterior border, one superior, the other anterior. Corresponding faces on the navicular bone of various rodents are in contact with the entocuneiform distally, and with an additional internal tarsal bone proximally, the internal navicular. These facets, therefore, indicate the existence of a hallux, but perhaps of a rudimental one. The mesocuneiform is smaller than the ectocuneiform, having much less transverse and little less longitudinal diameters. Neither are produced posterior to the proximal ends of their corresponding metatarsals. The third and fourth phalanges are about equal in length, and a little exceed the second. Their distal extremities have a prominent inferior median keel. At this articulation there is an ovate sesamoid bone on each side. A single proximal phalange, probably belonging to the *A. inundata*, is very much depressed at the distal articular extremity indicating a rudimental digit, doubtless the hallux, similar to that found in several existing allied genera.

The hinder foot of *Amblyrhiza* then is rather short, furnished with four developed digits and a rudimental hallux, and was probably plantigrade. The lack of tibial crest indicates that the knee was not constantly maintained in a flexed position. The immense trochanter indicates great power of extension of the femur, but whether this extension was effective in running or kicking is uncertain. The absence of tibial crest and the shortness of the foot militate against the supposition that these animals possessed powers of leaping, and their swimming powers would be impaired by the same structural characters.

Affinities.—This genus clearly enters Prof. Brandt's division of the *Rodentia* which he terms the *Hystricomorpha*. The evidence is seen primarily in the free fibula and in the development of the angular portion of the mandible on the external side of the incisive alveolus. The small coronoid process and the generic characters add to the weight of the evidence. Mr. E. R. Alston has recently published a very valuable *résumé* of the characters of the subdivisions of the

Rodentia, including the genera. He divides the *Hystricomorpha* into numerous families, some of which at least appear to the writer to rest on rather slender bases. In the comparison with *Amblyrhiza*, the *Hystricidæ* and *Dasyproctidæ* may be dismissed from the fact that their molars are not divided transversely by laminæ of enamel. The comparison is with the *Chinchillidæ* and *Caviidæ*. The molar dentition is that of the former family, and the absence of a masseteric ridge separates it from the genera arranged by Mr. Alston under the *Caviidæ*, although I cannot perceive that such a character should define a family group. The incisors of both these groups are called by Mr. Alston "short." I have shown those of *Amblyrhiza* to be very long, as in the *Dasyproctidæ*; nevertheless, their transverse section and sculpture are much as in the genus *Lagidium*. The affinities of this form are then near to types now existing on the South American Continent, but it presents characters which show that it cannot be referred to any existing genus. The extinct *Archæomys*, Laiz., Par., of the French Miocene, resembles it in the constitution of its molar teeth, but whether in other respects or not I cannot ascertain. There is one less dentinal column in the posterior superior molar of *Archæomys*, so that the formula for those teeth reads $\frac{5}{5}$ in *Amblyrhiza* and $\frac{4}{5}$ in *Archæomys*.

The only other extinct form with which it is necessary to make comparison is the one called by D'Orbigny, *Megamys patagoniensis*. This species rests on a tibia found in Patagonia, and it is described and figured in the Voyage en Amérique Meridionale, Vol. III. Pl. VIII. figs. 4–8. Although the specimen was injured, enough remains to show that it belongs to a different species and genus from any of those herein enumerated. The differences are to be seen in the distal extremity. They are, first, the absence of the divergent internal malleolar process, which is so striking in *Amblyrhiza*; second, the large size of the internal cotylus of the tibia, indicating that the internal rotula of the astragalus is as large as the external one, instead of being much smaller. The deep fissure which separates the articular faces of the head of the tibia is similar to that seen in *Amblyrhiza*.

My friend, J. A. Allen, the distinguished zoologist, remarks in his article on the *Castoroididæ*:[1] "To the same group (*i. e.*, the *Castoroididæ*) are, however, probably referable the genera *Amblyrhiza* and *Loxomylus*, described from the bone caverns of Anguilla Island, West Indies." . . . "The molars as described and figured by Professor Cope greatly resemble those of *Castoroides*, having, in fact, the same structure, differing mainly in being somewhat smaller and in possessing a greater number of laminæ. . . . As the lower jaw and skull are thus far unknown in these genera, it is impossible to say whether their affinities are strictly with the *Chinchillidæ* or whether they are not most closely allied to *Castoroides*." The description of the lower jaw and other parts of the skull found in the preceding pages show conclusively that *Amblyrhiza* has little affinity with the *Castoroididæ*, although, as Mr. Allen remarks, there is close resemblance between the molar teeth of the two forms. *Castoroides* differs from *Amblyrhiza* in its generally *Sciuromor-*

---

[1] Final Report of the U. S. Geological Survey of Territories, vol. xi. p. 421.

phous characters. The inferior incisive alveolus does not project on the inner side of the ramus, and the angle is a continuation of the general plane which incloses the alveolus. The ascending ramus is much more elevated, rising abruptly from the horizontal ramus. The symphysis is short, and, like the beaver, was not probably invested by the hairy skin as in the Chinchilla and *Amblyrhiza*.

History.—The first notice of this genus was published by the writer in the Proceedings of the Philadelphia Academy for 1868, p. 313, in the following language. Prof. Cope . . . also exhibited bones and teeth of a large rodent from the cave deposits of Anguilla, one of the Virgin West India Islands. The characters observed were those of the genus Chinchilla, but the roots of the teeth were contracted and not so open as in many rodents, as though having a more limited period of growth, or perhaps like deciduous teeth, which are much reduced in number in most rodents. The species was nearly as large as the *Castoroides ohioensis* of North America, but had relatively smaller incisor teeth. The body was probably as large as that of the Virginia deer, and the limb bones as stout, as seen in portions of femora and other pieces preserved. He called the animal *Amblyrhiza inundata*, and thought that its discovery on so small an island, with others of like character, indicated that the Caribbean continent had not been submerged prior to the close of the Post-pliocene, and that its connection was with other Antilles, while a wide strait separated it from the then comparatively remote shores of North America.

Subsequently in the Proceedings of the American Philosophical Society for 1869, p. 183, I more fully defined the genus from more complete material received from Dr. van Rijgersma. I at the same time described the inferior molars, but on account of the great difference which they display as compared with the superior molars, I did not suspect that they belonged to the same genus and species as the latter. I then gave them other names, generic and specific, which are now withdrawn. In the same periodical for 1870, p. 608, I added the species *A. latidens*, and in the same of the following year (p. 102) I added some observations on the geographical relations of the genus, and described the species *A. quadrans*.

AMBLYRHIZA INUNDATA, Cope.

Proceedings of the American Philosophical Society, 1869, p. 183. *Loxomylus longidens*, Cope, loc. cit. 1869, p. 186. Plates IV. and V., figs. 1–3.

The remains of this large rodent were found in a mass of breccia, which was thrown out in the excavations made in a cavern in the Virgin Island of St. Martins, W. I. The remains, occurring in that most eastern region of the West Indian Zoological district, might be anticipated to have a special interest in connection with the history of the submergence of a once great continent. With this impression, the writer examined a quantity of the above breccia and cave deposit, which was brought to Philadelphia as a probably available phosphatic manure. It was found to be valueless for this purpose, and the only result of the outlay was the discovery of the subjects of the present memoir. Most of the fragments first described were dressed from a single block. There were in this, the extremity of

2   April, 1883.

a right femur with patella, shafts of various long bones, fragments of pelvis and maxillary bones, with three molars, and two partially complete, and other much broken incisors. The teeth were scattered among the bones, and are so related in size to most of them as to induce the belief that they all belong to the same animal. This is strengthened by the entire absence of bones or fragments which could be referred to any other animal.

Having requested Dr. van Rijgersma to make further search in the cave excavations, that gentleman made an especial expedition to Anguilla for that purpose. His efforts were rewarded by the recovery of ten or more teeth of the inferior series, and a number of the superior molars of two individuals with various bones. Teeth of five individuals of the species have probably been obtained in all. Dr. van Rijgersma on subsequent expeditions obtained the other species, *A. quadrans* and *A. latidens.*

Of the first individual, three *superior molars* are preserved, which present four dentinal columns. These columns are transverse, the first, which I assume to be anterior, transverse; the second, the longest; the third, shortened inwardly and slightly curved round the very small fourth, which occupies a postero-external angle of the crown. All are separated by rather thick enamel laminae. The form of the crown of the largest presents two sides of a square anteriorly and externally; the inner side bilobed in correspondence with the two anterior columns; the posterior strongly convex backwards and outwards. The other, similar molar, differs in the posterior outline being more nearly transverse, and the anterior and interior outlines being united by a continuous curve. The large portion of the third tooth preserved is perhaps the external; it is part of a nearly regular transverse oval.

The first described molar is strongly curved posteriorly, and its diameter narrows regularly to the contracted base; there is a shallow groove at the junction of the anterior enamel lamina with the inner wall. This groove is much more strongly marked in the second described, but ceases before attaining the contracted extremity. The shank of the tooth is less curved than in the other. The contraction is less graded than in the first, but is strongly marked at the base, where the pulp cavity is not wider than one of the columns.

				Lines.
Length of anterior face, No. 1 (on curve)	.	.	.	14 3
Diameter of crown (longitudinal)	.	.	.	6
" " (transverse)	.	.	.	5.7
" of root (longitudinal)	.	.	.	4
Length of anterior face, No. 2	.	.	.	14.3
Diameter of shank (longitudinal)	.	.	.	6
" " (transverse)	.	.	.	5

A portion of one of the inferior incisors of some forty-six lines in length, and another shorter piece, furnish characters of the species and genus. The inner face of the tooth is plane, and at right angles to the anterior; the outer is rounded obliquely inwards; the inner face is broad and not prolonged; the curve of the tooth is in one plane, and the depth is about equal to the width. A narrow fold of the enamel embraces the anterior border of the inner and outer faces; it is folded back at a right angle within.

A section of all except the terminal teeth is an oblique rhomboid, the longitudinal diameter being but little greater than the transverse. A single terminal tooth (either superior posterior or anterior inferior) is narrowed in the terminal column. All the teeth possess one longitudinal groove on one side, and two on the other, which are covered but not obliterated by the cement layer. The teeth, though much straighter and more slender than those of the superior series, yet possess a slight lateral, though no anteroposterior curvature; those of the upper and lower series curving in opposite directions.

			Inches.
Length of a median molar	. . . .	.	1.7
Anteroposterior diameter (oblique)	. . .	.	.52
Transverse " (both of crown)	. .	.	.43
" " (terminal molar)	. .	.	.39
Longitudinal diameter of terminal molar	. .	.	.56

The inner face of the incisor tooth is plane and at right angles to the anterior; the outer is rounded obliquely inwards; the inner face is broad and not prolonged, the curve of the tooth is in one plane, and the depth is about equal to the width. A narrow fold of the enamel embraces the anterior border of the inner and outer faces; it is folded back at a right angle within, and with a truncate angle without. The enamel is sculptured into numerous close, fine longitudinal grooves, which do not inosculate. The separating ridges number fourteen near the middle of the tooth, those near the borders being the strongest. One, strongest of all, is on the external turn of the enamel, and near it numerous interrupted ridges have a slightly oblique direction.

The incisors are, as in the modern representatives of the *Amblyrhiza*, of more slender proportions than in the beavers, *Arctomys*, and other rodents, and their extinct predecessors. They are therefore relatively less stout than those of the *Castoroides*. Their sculpture is quite similar to that seen in the *Lagidium* and other chinchillas.

			Lines.
Width anteriorly	. . . . .	.	6
Depth	. . . .	.	5.7

The inferior molars differ much from those of the superior series. They are straight, prismatic, and composed of three dentinal columns, one of which is incurved, but none closed at the base. The triturating surface is very oblique in the vertical direction, indicating the greater elevation of the teeth at one extremity of the series than the other. A horizontal obliquity of the dentinal columns is produced by their lateral displacement. Enamel plates but slightly curved.

Extremities.—I describe under the head of this species those bones of adult individuals of smaller dimensions which the different lots contain. Some of these came associated with teeth of *A. inundata* without admixture of larger bones; and most of the larger ones came to hand with the jaws and teeth of the *A. latidens* and of the *A. quadrans.* In other lots both sizes were mixed.

A femur distinguished by its bright red color came with teeth of *A. inundata* of the same color. The triangular form of the section of the shaft disappears a short distance below the little trochanter, and is replaced by a suboval one. The measurements are as follows:—

		M.
Diameter of base of great trochanter .	. . . .	.042
Vertical diameter of neck	. . . .	.023
" " of head	. . . .	.032
Length of head and neck	. . . .	.033
Diameter of shaft at middle { transverse	. . . .	.031
{ antero-posterior	. . . .	.027

A proximal phalange of the inner digit of the posterior foot, or the hallux, accompanied this femur, and presents the same color. The obliquity of its proximal articular cotylus demonstrates that its position in life is subhorizontal. There are two prominent proximal inferior tuberosities. The distal extremity is depressed, so that the planes of the superior and inferior surfaces form an acute angle with each other. In the middle of this extremity there is a rudiment of trochlear keel. The lateral ligamentous fossæ are distinct but small, and are bounded by a low tuberosity proximally.

		M.
Length of phalange	. . . . .	.026
Vertical proximal diameter { total	. . . .	.013
{ of cotylus	. . . .	.008
Transverse proximal diameter .	. . . .	.014
" distal "	. . . .	.008

This phalange I originally described as distal, an error I now correct. The teeth which accompanied it are figured on Plate II. fig. 6.

In a second individual the bones are earth-colored, and were not associated with teeth. The middle of the shaft of the humerus is extended in antero-posterior diameter, in consequence of the existence of an anterior median ridge. The inner epicondyle is more distinctly developed than in specimens described under *A. latidens*. The crest or anterior ridge of the accompanying tibia is less prominent than in those of the larger animals. The external glenoid surface overhangs very far, and is supported by a lateral rib of no great prominence. The middle of the shaft is strongly compressed so that the transverse diameter is small. The distal extremity, perhaps of the same tibia, is characterized by the great internal projection of the processes, and absence of the ridges which rise from them. The tendinous groove is very slightly roofed. The measurements of these bones are as follows:—

		M.
Diameter of shaft of humerus { fore and aft .	. .	.023
{ transverse	. .	.021
" of distal extremity { fore and aft	. .	.021
{ transverse	. .	.045
" of head of tibia { fore and aft	. .	.044
{ transverse .	. .	.062
" of shaft of tibia { fore and aft	. .	.025
{ transverse .	. .	.019
" of extremity of tibia { fore and aft .	. .	.023
{ transverse	. .	.034
Transverse extent of astragular facets .	. .	.028

Associated with these fragments and having the same brown color, is a portion of the hind foot, which has furnished characters for the generic description. This specimen is large enough to accommodate the astragalus which accompanies the bones referred to *A. latidens*, and I, therefore, am not sure to which species it belongs. The exposed front face of the navicular bone is nearly twice as wide as long, while that of the ectocuneiform is subquadrate. The tarsal facets of the metatarsal bones are very oblique, indicating that the axes of the latter are directed more anteriorly than that of the tarsus, causing a constant flexure of the foot. The posterior angle of the second metatarsus is produced proximally behind the mesocuneiform, while that of the third metatarsal projects behind the ectocuneiform in an obtuse tuberosity. The shafts of the second, third, and fourth metatarsals are proximally compressed, and medially depressed. The fourth metatarsal articulates with the cuboid proximal to the articulation of the second with the ectocuneiform, a character to be seen in a less degree in the articulation of the second metatarsal with the mesocuneiform. It, therefore, follows that the line of tarsal junction of the metatarsals is a convex semicircle.

DIMENSIONS OF HIND FOOT.

			M.
Diameters of navicular	longest (oblique)	. .	.032
	anteroposterior	. .	.024
	transverse (front)	. .	.021
	longitudinal (front)	. .	.014
" of mesocuneiform	anteroposterior	. .	.017
	transverse	. .	.010
	longitudinal	. .	.010
" of ectocuneiform	anteroposterior	. .	.020
	transverse	. .	.015
	longitudinal	. .	.015
Transverse diameter of I., II., and III. metatarsals		. .	.037
" " of II. metatarsals		. .	.011
" " of III. "		. .	.011
Length of metatarsal, No. I.	. .	. .	.037
" " No. II.	. .	. .	.043
" " No. III.	. .	. .	.044

The proximal portions of three scapulæ accompanied the head of a humerus, and the proximal halves of ulnæ and radius of other adult individuals of the smaller general proportions. These display well excavated supra- and infra-spinous fossæ, the latter the most pronounced. The plate of the scapula expands rather widely immediately beyond the short neck; the glenoid cavity is a transverse oval, narrowed and produced at the coracoid extremity. The head of the humerus proper is subequal in its diameters. A section of the ulna just beyond the coronoid process presents five angles, one superior, one inferior, two interior, and one exterior. The upper and lower interior planes thus formed are the narrowest; the two superior and the exterior soon fuse into a convex face. The coronoid humeral face is much narrower than the olecranon, whose rim extends downwards on the inner side to the middle of the shaft. The olecranon is flat below, and as deep as wide at the end. The head of the radius expands laterally from the neck. Its articular face narrows outwards, being reduced by a bevel of the inner part of the superior edge. Its face is a little concave, and is not sigmoid in section as in some mammals.

MEASUREMENTS OF FORE LIMB.

		M.
Diameter of glenoid face of scapula { vertical .		.020
{ transverse		.035
Transverse diameter of neck of scapula	.	.035
Length of coracoid process (horizontal)	.	.015
Elevation of spine of scapula	.	.010
Diameter of head of humerus { transverse .		.033
{ fore and aft .		.028
" of ulna at middle { transverse	.	.014
{ vertical	.	.024
Depth of ulna at coronoid process	.	.033
" of olecranon	.	.022
Length "	.	.032
Width "	.	.019
Diameter of radius at middle { vertical	.	.010
{ transverse	.	.018
" of head of radius { vertical	.	.016
{ transverse	.	.023

AMBLYRHIZA QUADRANS, Cope.

Loxomylus quadrans, Cope, Proceedings Amer. Philosoph. Soc 1871, 102.

This species is represented by teeth intermediate between the *A. inundata* and *A. latidens* in size. It was originally established on a first superior molar which is worn down so far as to have lost the characteristic curvature, and to be nearly straight in the shaft. Its section is nearly a quarter of a circle, a little longer anteroposteriorly than transversely, not so elongate as the corresponding tooth of *A. inundata*, but more as in *A. latidens*. The dentinal columns are oblique in the transverse direction, the second being most extended, the third next, the first a little shorter, and the fourth shortest, and transversely triangular in form. The internal and posterior borders are nearly straight, and form a little more than a right angle with each other. The sides of the shaft are not grooved, except a shallow interruption between the first and second plates on the external aspect.

					M.
Length of shaft preserved	.	.	.	.	.023
Diameter of grinding face { anteroposterior	.	.	.	.	.015
{ transverse	.	.	.	.	.014

The posterior portions of both mandibular rami of one individual support teeth which differ from the corresponding ones of *A. latidens*, and which agree in proportions with the tooth above described. I refer them provisionally to the same species.

The left ramus exhibits the alveoli of all the teeth. As in the other species they are straight, and directed obliquely upwards and forwards, and are not oblique laterally. The third and fourth are a little longer than wide, and their alveolar septa are a little oblique outwards and backwards. The second alveolus is smaller and relatively wider than the others.

MEASUREMENTS OF TEETH.

					M.
Long. diameter of fourth molar	.	.		.	. .017
Transverse " "	.	.		.	. .014
Long. diameter of third alveolus	.	.		.	. .016
Transverse " "	.	.	.	.	. .014
Long. diameter of second alveolus	.	.	.	.	. .013
Transverse " "	.	.	.	.	. .014

The form of the crown of the last molar differs from the corresponding tooth of the *A. latidens* in its greater anteroposterior than transverse diameter; measurements which are equal in the latter species. This is due to the greater anterior projection of the middle dentinal column on the outer side, and the consequent obliquity and transverse shortening of the anterior column in *A. quadrans*. The section of the tooth in this species is 8-shaped; in *A. latidens* nearly quadrant-shaped. The lateral groove in *A. quadrans* is on the inner side at the front of the middle column; on the outer side, behind it.

The coronoid process begins opposite the third column of the last molar, and its base terminates opposite to the dental foramen. The incisive alveolus reaches to this foramen, which is at the base of the flat front of the ascending condylar portion of the ramus. The external fossa of the ramus is an inch below the alveolar border, and is opposite the septum between the second and third alveoli. There is a distinct masseteric fossa, as indicated by a portion of its anterior part. The anterior border extends to a point one-half inch behind that below the posterior edge of the fourth molar.

MEASUREMENTS.

					M.
Length of inferior molar series (anterior part of m. 1 estimated)			.	.	.060
" from m. 4 to mental foramen	.	.	.	.	. .025
" of coronoid process from base	.	.	.	.	. .014
" of base of coronoid process	.	.	.	.	. .036
Width of base of condyloid branch at foramen		.	.	.	. .020
Diameter of dental foramen	.	.	.	.	. .020
" of external fossa	.	.	.	.	. .013

I mention here a superior incisor tooth, which is in dimensions intermediate between the largest and smallest of the genus. I do not know whether it represents a distinct species from those possessing the usual form of incisor, or whether its peculiar character is abnormal. With the surface of other incisors it is openly grooved along the middle in a perfectly regular manner. The surface on the inner side of the groove is plain; on the outer side, the front is convex. Width of front m. 014; depth of shaft .015.

AMBLYRHIZA LATIDENS, Cope.

Loxomylus latidens, Proceed. Amer. Philos. Soc. 1870, p. 608.

A number of bones and teeth from Simson's Bay came associated in one package, and agree in having a paler color than the specimens contained in other packages. Among them is a portion of the right maxillary bone of an *Amblyrhiza*, containing the last two maxillary teeth, and having attached the adjacent portions of the palatine bone. The maxillary bone projects between the last molar tooth a distance equal to more than half the diameter of the latter. The palatine follows, presenting some surface continuous with the external face of the maxillary; its maxillary suture pursues a nearly straight course forwards a short distance inside the crowns of the teeth. It begins to curve inwards opposite the middle of the penultimate molar, but continues some distance further forwards. The teeth nearly resemble those already referred to the *A. inundata*, but are considerably larger. The first molar of the left maxillary which accompanies the other pieces, is also materially different in form from the corresponding tooth of that species.

The first superior molar includes four dentinal columns, of which the posterior is much the smallest, and is situated on the inner angle of the posterior face of the tooth. The enamel bands are directed a little backwards as well as outwards, and the one between the first and second columns has an inflexion near its inner extremity. The anterior external enamel plate is regularly convex in section and oblique in position, striking the dividing plate which follows it at a point on the outer side of the crown which marks the middle of the long diameter. In the *A. inundata* the corresponding tooth is not only very much smaller but is relatively narrower in transverse diameter, and the anterior column does not extend so far backwards on the outer side. This tooth is, in the *A. latidens*, as large in all its diameters as the third or fourth molars.

The second molar is wanting from the specimen, but its alveolus shows that its posterior wide columns project further inwards than in either of the teeth which follow. The grinding surface of the third superior molar is straight on the inner side and convex on the outer through the prominence of the second and third columns; the fourth is small and internal. The anterior column of the fourth molar is more transverse than the others or than those of the third, which are oblique. The inner side of the grinding face of the tooth is longer than the outer, owing to the presence of the small fifth column.

A pair of incisors with both extremities wanting, which are held together by the matrix, accompanied the above specimens. They belong to the upper jaw, and as they are of appropriate size to the superior molars just described, I suspect that they belong to the same animal. Their section is triangular, with the external anterior margin projecting and curved with an incurved border of the enamel sheet. The surface of the latter is marked with numerous irregularly spaced longitudinal grooves, excepting for a narrow space along the inner margin, which is smooth.

MEASUREMENTS.

		M.
Length of first molar .	. . .	.033
Diameter of grinding face of first molar { fore and aft		.018
{ transverse		.016
" " " of third molar { fore and aft		.015
{ transverse		.015
" " " of fourth molar { fore and aft		.017
{ transverse		.014
" of shaft of superior incisor { fore and aft .		.017
{ transverse .		.017

The larger number of specimens received from Dr. van Rijgersma are of a dull rusty-brown hue, and were doubtless derived from one and the same locality, and were found at or near the same time. They include the bones and teeth of the largest specimens of the genus *Amblyrhiza*, while most of the smaller individuals were received at other times, and from differently colored deposits. The mandibles and teeth of two different species are included in this lot, and one of them is probably the *A. latidens*, but which one is to be collated with the maxillary bone above described is not certain. As the maxillary teeth of *A. latidens* differ in the specific character of a greater width from those of *A. inundata*, it is probable that the inferior molars which bear the same relation to those of *A. inundata*, are the ones to be described here. Those which present other differences, which are less marked, are referred to another species under the name of *Amblyrhiza quadrans*.

Of mandibles referable to the *A. latidens*, there are both sides of one individual in a more or less mutilated condition, one only presenting the molar series complete. Seen from the outer side, the summits of all the inferior molars rise an equal distance above the alveolar border, the insignificant excess in height of the first tooth being only individual. But on the inner side the alveolar edge descends forwards so as to leave the first molar projecting to four times the elevation of the fourth. Viewed from above, the molars successively narrow forwards, the first being only half as wide as the fourth. The direction of the transverse enamel bands becomes more oblique forwards externally, so that the anterior column of the first molar stands at an angle of 45° to the long axis of the series. Those of the fourth molar are moderately oblique, and the middle column in the third and fourth molars has the greatest transverse extent. The inner posterior angle marks the middle of the basis of the small coronoid process. The apex of the latter is recurved, and the direction of its plane is 45° from that of the mandible. The inner alveolar border is continued as a ridge for a short distance behind the last molar, and then descends and disappears at the distal foramen. This foramen opens on a horizontal surface or plane which is bounded without by the external border of the ascending limb, and within by the strong ridge which connects that portion with the prominent alveolar sheath of the inferior incisor. The inner face of the jaw is gently concave to the incisor alveolar ridge; the external face is marked by a fossa, about as wide as the diameter of a molar tooth 25^{mm} below the alveolar border, and opposite the third and part of the second teeth.

MEASUREMENTS.

		M.
Length of series of molars (No. 1) .	. . . .	.058
" of posterior molar .	. . . .	.047
Projection of posterior above alveolar edge	. .	.005
Diameter of fourth molar { fore and aft	.	.016
{ transverse	.	.016
" of third molar { fore and aft	.	.013
{ transverse .	.	.015
" of second molar { fore and aft	.	.012
{ transverse .	.	.011
" of first molar { fore and aft	.	.015
{ transverse	.	.011
Elevation of first molar on inner side .	.	.015

The premaxillary bones with incisor teeth of several individuals accompany the mandibles just described, and some of them probably belong to this species, and others to the *A. quadrans*. I cannot detect any difference in the two which are best preserved, and which correspond to the two individuals represented by molar series, and symphyses with inferior incisors. The section of the produced pair of premaxillary bones is triangular near the middle. That of the lateral walls is greatly convex, and the median line below is keeled for a short distance. Immediately behind the keel is a narrow fossa. The section near the base of the incisors is subquadrate, and there is a deep longitudinal median fossa just behind the base of the teeth, whose fundus appears to receive a foramen.

The symphyses above mentioned are long and acuminate. The union of the inferior surface continues posterior to the point of divergence of the dental ridges, forming the bottom of a large fossa. The section of the anterior or beak-like portion of the symphysis is subtriangular, the sides approaching each other upwards, but separated by a broadly obtuse median ridge. Near the middle of the length a slight enlargement is visible in both the specimens. The inner face of the incisor teeth is more strongly convex than the external.

MEASUREMENTS.

						M.
Length of beak of symphysis .		.	.	.	.	.090
Diameter of beak at middle	vertical	.	.	.	.	.025
	transverse	.	.	.	.	.023
" of inferior incisors	vertical	.	.	.	.	.014
	transverse	.	.	.	.	.014

The most complete mandible includes the right ramus with all the alveoli, the entire symphysis with the right incisor tooth in place. The molar alveoli are empty. The posterior alveolus is as wide as long, as in the typical specimen, and is in form about a quarter of a circle. The third and second alveoli are similar in form, but a little more quadrate, and the second is materially the smallest of all. The third is much longer and a little wider than the second. The symphyseal beak is more attenuated than in the specimens just described, and its superior middle line is more angular, giving a roof-shaped section. The incisor tooth is exposed on its external side for almost the entire length of this rostrum, to a point 17ᵐᵐ anterior to the divarication of the alveolar ridges. The cutting edge of the incisor is transverse. The lateral fossa is below the second and part of the first molar. This is anterior to the position which it occupies in the mandibles described under *A. latidens* and *A. quadrans*. This fact, in connection with the slight difference in the form of the symphyseal rostrum, has led me to question the propriety of the reference of the animal to the *A. latidens*. The fourth molar evidently has the form characteristic of that species, rather than that of the *A. quadrans*.

MEASUREMENTS.

	M.
Length from apex of symphysis to posterior base of coronoid process	.160
" of symphyseal rostrum	.080
" from base of rostrum to the first molar	.012
" of free inner face of incisor	.038
" " outer " "	.110
" of alveolus of first molar	.018
" " of second molar	.011
" " of fourth molar	.016
Width " " "	.016
" between rami at fourth molars	.050

The mandibles, with teeth of the two larger individuals last described above, were sent by Dr. van Rijgersma in connection with numerous bones of the skeleton, which, in color, mineral character, and dimensions, are so appropriate to them, that I do not hesitate to believe that they belong to the same individuals, excepting in so far as they include portions of the skeleton of the *A. quadrans.* Of these the vertebræ first claim attention.

The atlas is depressed in form, and the condylar fossæ are divergent upwards. The odontoid fossa viewed from the front has only half the diameter of the neural arch. The transverse processes expand backwards and outwards, their postero-external angle being behind the plane of the facets for the axis, and the border being a regular curve from a short base behind the margin of the condyloid fossa. The axial facets are wide oblique ovals separated below and bounded by a deep groove at the base of the neural arch. The neural spine is lost.

The neural canal of the axis is subquadrate, and a little wider than deep. The neural spine is strongly keeled in front, the inferior extremity of the keel projecting beyond the arch. The posterior face of the spine is subconcave. The posterior zygopophyses are well developed. The posterior articular face of the cerebrum is a hexagon, a little wider than deep, and its vertical plane is directed a little forwards above. The condyles for the atlas descend to the plane of the inferior surface.

DIMENSIONS OF CERVICAL VERTEBRÆ.

		M.
Length of atlas below	(least (median)	.010
	(greatest	.025
Width of atlas	{ anteriorly	.059
	{ posteriorly	.108
Length of transverse process behind		.032
Depth of " " at base		.020
Width of neural canal		.025
Vertical diameter of facet for axis		.016
Length of centrum of axis (without odontoid)		.024
" of odontoid process		.014
Diameter of posterior face of centrum	(vertical	.017
	(transverse	.025
Greatest width of centrum		.040
Interior elevation of neural arch		.018
Diameter of neural spine	(transverse	.020
	(antero-posterior	.018

The centra of the lumbar vertebræ are strongly constricted at the middle, and they display an arterial foramen near the middle of each side; their inferior surface is regularly convex, and without keel. One of the articular faces has a slight tendency to convexity. Neural canal wide. The proximal extremity of the first sacral centrum is plane, and not depressed nor concave. The inferior faces of the sacral vertebræ are slightly concave in the anteroposterior direction. The centrum of one of the canals is oblique in the vertical plane of one of its extremities. Its inferior surface is strongly convex, the superior moderately so. The base of the diapophysis is oblique.

DIMENSIONS OF LUMBAR AND SACRAL VERTEBRÆ, ETC.

			M.
Length centrum No. 1 .		.	.036
Diameter anterior articular face	{ vertical	.	.038
	{ transverse	.	.038
" posterior face	{ vertical	.	.041
	{ transverse	.	.046
Length centrum No. 2	.	.	.045
Diameter anterior articular face	{ vertical	.	.039
	{ transverse	.	.044
" posterior articular face	{ vertical	.	.040
	{ transverse	.	.054
Length three sacral centra	.	.	.113
" of first sacral centrum	.	.	.052
Diameter of articular face of first sacral	{ vertical		.042
	{ transverse		.048
Length of third sacral centrum	.	.	.026
Diameter of third sacral centrum in front	{ vertical	.	.017
	{ transverse		.026
Length of centrum of a caudal	.	.	.022
Diameter of articular face of a caudal	{ vertical	.	.023
	{ transverse	.	.024

No scapula certainly referable to this species has been preserved. The proximal and distal extremities of the humerus differ from those of the A. *inundata* in their superior size. The supracondylar fossa is not perforate. The fossa of the great trochanter of the femur is only excavated in the inferior half, its lower border being opposite the middle of the neck of the femur. Its superior portion is the deepest, and the fundus corresponds to a projection of the opposite border of the external face. The external face of the femur is flat to as far below the little trochanter as the length of the great one, when it passes by a regular curve into the posterior face. The angular ridge bounding the anterior face continues further down. The anterior face of the femur is well defined by lateral parallel ridge-angles, and maintains a nearly equal width from its middle to near the proximal extremity of the great trochanter. Near the middle (where my specimens are broken off) it is gently convex, but it soon becomes flat, and is slightly concave at the base of the great trochanter. The extremity of the latter is a wide continuation of the external face of the shaft, rising obliquely inwards. Its borders are roughened and at some points overhang, furnishing points of muscular insertion. It projects beyond the neck a distance equal to 1.25 transverse diameters of its base. The *fossa ligamenti teris* is small but deep, and far removed from the edge of the head, which is not connected with it by a groove. The condyles of the femur are rather divergent. The condylar surfaces are continuous with the patellar. The external bounding angle of the latter is a little more elevated than the internal. The latter bounds a trochlear bevel which extends band-like along its inner side. This looks as though the patella possessed a corresponding bevel which would include an angle with its remaining inferior surface.

The proximal portion of the tibia is characterized by the depth and width of the longitudinal external fossa, which is bounded by the crest. The proximal articular faces are not concave. The posterior face of this part of the shaft is narrow. The distal portion of the shaft is marked on its external aspect by a narrow groove which runs behind the fibular ridge, terminating where the latter begins to expand. The angle rising from the anterior malleolar tuberosity becomes the internal edge of the shaft, while the one which arises from the posterior one disappears on the posterior face. The anterior and posterior edges of the external glenoid cavity are equally produced, and the external and internal borders have the same relation. The internal superior facet of the tibial articulation of the astragalus is half as long and half as wide as the internal; the area of the internal lateral face is still smaller. There is a shallow fossa at the neck on the outer side. The navicular facet is subround, and deeper than wide; the facet for the additional proximal tarsal bone is subtriangular with the apex posterior; its surface is moderately convex.

MEASUREMENTS.

			M.
Diameter of distal extremity of humerus	fore and aft		.029
	transverse		.062
" of acetabulum	vertical		.041
	transverse		.038
" of neck of ilium			.038
" of shaft of femur near middle	fore and aft		.029
	transverse		.043
" of base of great trochanter	fore and aft		.034
	transverse		.045
Length of great trochanter			.065
" of head and neck			.036
Diameter of head of femur			.036
" of neck of femur			.026
" of condyles of femur (transverse)			.085
Fore and aft diameter of femur at patella			.052
Diameter of head of tibia	fore and aft		.060
	transverse		.077
Width of external articular face			.030
" of internal "			.035
" of posterior plane of shaft			.030
Diameter of inferior part of shaft	fore and aft		.020
	transverse		.024
" of distal extremity	anteroposterior		.035
	transverse		.049
Transverse diameter of trochlear face			.031

					M.
Width of astragalus	{ posteriorly	.	.	.	. .044
	{ medially				. .030
" of external superior face		.	.	.	.017
" of internal "		.	.	.	. .009
" of internal face .		.	.	.	. .009
Depth at internal border		.	.	.	. .044
" of internal facet of head		.	.	.	.012
Diameter of navicular facet	{ vertical .	.	.	.	. .023
	{ transverse	.	.	.	. .018

Portions of a sacrum and a lumbar vertebra of another individual of the size of this species present some additional characteristics. Thus the anterior margin of the middle of the neural arch in the first sacral is produced into a pointed process which separates two fossæ. The anterior zygapophyses in front of these are well developed, directing their articular faces upwards and outwards at an angle of 45°. The diapophysis of the lumbar arises from the summit of the neurapophysis, and is directed forwards at an angle of 45°. The section of the shaft is triangular and the apex is acuminate. Its length is .030; vertical diameter of neural canal .024. The anterior zygapophysis is produced upwards and a little outwards; its face is nearly vertical.

ARTIODACTYLE.

The following characters are derived from the right maxillary bone of a young animal which supports the temporary dentition, and from the bones of two individuals, one adult and the other young, probably the one to which the maxillary bone pertained. I give the full characters for the benefit of those who have the opportunity of studying the numerous species of the order found in the Old World.

Temporary superior dental series of the maxillary bone, Pm. 1; M. 3; other teeth unknown. Last premolar with an external subacute column, and an internal short column, which are united posteriorly, forming a rather wide grinding face, which has an external cutting edge; no single anterior column. True molars with four usual crescents, whose anterior horns are the inferior extremities of vertical ribs. The lakes are without cementum, and are united at their adjacent angles for some distance into the crown. The crowns of the molars are moderately elongate, and the roots of the anterior teeth are rather long.

The temporary premolar of this genus is more simple than the last of the permanent series in *Auchenia*, but presents a more complete inner crescent than does the penultimate premolar in that genus. Its inner crescent is better developed than that of the genus *Camelus*, since it is united with the external one at both extremities. The animal is apparently a member of the *Boridæ*.

Attached to the entire maxillary bone are the malar and palatine bones, in complete preservation. These bones display some interesting relations which I now describe. The premaxillary bone evidently reached the nasals in the complete skull. The palate is deeply emarginate posteriorly. The emargination between the maxillary and the *processus pyramidalis* of the palatine is profound. The palatine bone extends forwards on the middle line to opposite the middle of the dental series. In front of the processus pyramidalis, and separated from it by a deep fissure, the thin lamina perpendicularis rises. Its anteroposterior diameter is greater above than below. Its anterior margin articulates by squamosal suture with the interior side of the superior alveolar wall of the maxillary bone. This wall is so strongly convex as to divide the maxillary sinus into two fossæ, an internal large, and external small. The latter is bounded internally by the wall of the infraorbital canal, which extends along the summit of the alveolar arch. The nasal roof of the antrum is thin, and only extends as far posteriorly as the middle of the first true molar. The posterior extremity of the malar bone is deeply notched, and was connected with the frontal, inclosing the orbit behind.

The palate is rather wide, and the molar series is curved or convex outwards. The posterior external column of the last true molar is half as wide again, in anteroposterior diameter, as the corresponding anterior column. The vertical ribs are most prominent in the second and third true molars, the anterior being obsolete on the first true molar, and wanting on the premolar. In all the molars the anterior column has a median external rib, which is strong on the first true molar, less marked on the other teeth, and wanting on the posterior column in all. The two inner columns are separated by a vertical fissure to a considerable depth in the second true molar, but not in the first. There is a slight contraction of the anterior lake of the first true molar caused by a low rib on the posterior part of its internal enamel wall.

The form of the face is short and high, the facial plate of the maxillary rising above the level of the lachrymomalar suture. The external foramen of the infraorbital canal is large, and issues above the posterior part of the premolar. There is no lateral facial fossa. The inferior orbital border is strongly flared outwards. The posterior palatal notch enters as far as the anterior margin of the last molar, and the external pterygo-maxillary notch is not quite so deep. The palato-maxillary suture leaves the alveolar border at the front base of the last molar.

MEASUREMENTS.

		M.
Length of entire molar series	. . .	.047
" of true molars.	. . .	.039
Diameter of premolar { anteroposterior	. . .	.0075
{ transverse .	. . .	.0055
" of first molar { anteroposterior	. . .	.0120
{ transverse .	. . .	.0055
" of second molar { anteroposterior	. . .	.013
{ transverse	. . .	.006
Width of palate in front of premolar .	. . .	.014
" " at second true molar	. . .	.030

Associated with the cranial mass already mentioned as representing this species, there are several bones of the skeleton of an immature mammal, which are appropriate to it in size and proportions. These are an axis and a dorsal vertebra, humeri, radius, pubis and ischium, femur and cannon bones, all without epiphyses.

I now describe the other bones already mentioned. There is no sinus in the anterior border of the obturator foramen. The ischium has a flat section, with the long axis subtransverse. The section of the pectineal part of the pubis is a wide oval. There is a prominent transverse ischiatic tuberosity. The axis is moderately elongate, and the summit of its neural spine is horizontal. The anterior border of the neural arch is pierced by a large foramen on each side. The diapophyses are small and originate as a keel on the posterior half on each side, which continues into a short acute posteriorly directed process. The dorsal vertebra is short and has an opisthocœlian centrum. The diapophyses are short and with down-looking articular surfaces, and the neural spine is narrow and high.

The distal fossæ of the humerus do not communicate; the olecranar is deep. The distal extremity is more transversely expanded than in *Procamelus occidentalis*. The radius is flat, and with convex upper surface. The shaft is slightly arched upwards, and the section of each extremity is a transverse oval of subequal long diameters; the distal end has the greater vertical diameter. The shaft of the femur is not very long and is subcylindric. There is a slight anteroposterior curvature. The great trochanter is large, and is strongly recurved posteriorly. The posterior cannon bones are rather stout, and the section at the middle is subquadrate with rounded sides. Of the proximal facets the ectocuneiform is the largest. The anterior cuboid is a quarter of a circle, and is separated from the narrow posterior cuboid by a wide groove. The mesocuneiform is larger than the posterior cuboid, and is rather wide; its posterior border forms a protuberance. The inner extremity of the posterior cuboid is the most elevated part of the proximal end.

The distal fissure does not extend above the bases of the epiphyses.

MEASUREMENTS (WITHOUT EPIPHYSES).

				M.
Length of axis on centrum (laterally)	.	.	.	.023
Total elevation of axis	.	.	.	.030
Anterior diameter of centrum of axis { transverse	.	.	.	.030
{ vertical	.	.	.	.008
Posterior " " { transverse	.	.	.	.027
{ vertical	.	.	.	.011
Total elevation of anterior dorsal	.	.	.	.050
Elevation of neural arch	.	.	.	.012
" of spine	.	.	.	.032
Length of centrum	.	.	.	.010
" of humerus	.	.	.	.081
Transverse diameter proximally	.	.	.	.021
" " distally	.	.	.	.023
" " of shaft	.	.	.	.009
Length of radius (chord)	.	.	.	.078
Transverse diameter proximally	.	.	.	.020
" " medially	.	.	.	.011
" " distally	.	.	.	.021
Length of femur	.	.	.	.099
Transverse diameter at trochanter	.	.	.	.016
" " medially	.	.	.	.009
" " distally	.	.	.	.020

			M.
Length of cannon bone .	. .	. .	.065
Transverse diameter proximally	.	. .	.016
" " medially .	.	. .	.009
" " distally .	.	. .	.019
Length of ischium .	. .	. .	.047
Width medially	. .	. .	.010
Length of pubis to symphysis	.	. .	.012
Greatest diameter of symphysis	.	. .	.007

Several bones of a second individual accompanied those above described, which are of larger size and present adult characters. They include mutilated scapula and humerus, a nearly perfect ulno-radius, a metacarpal cannon bone, and a phalange.

As compared with those of a rather small *Oris aries*, these bones are smaller, and present the following differences: The glenoid cavity of the scapula presents a straighter or less convex internal border. The coracoid process is a little longer; the base of the spine is nearer the anterior edge. The form of the humeral shaft is similar, but the internal peduncle of the condyles has greater anteroposterior width. The extremities of the bone are lost. The radius scarcely differs from that of the sheep. I notice a little greater depression of the external part of the head, and a rather longer proximal interosseous foramen; the fossa behind the scaphoid facet is deeper. The proximal end of the metacarpal cannon bone has the same form as that of the sheep, but differs in the larger size of the facet for the *os trapeziomagnum*, which is also without the large interior emargination seen in the sheep. Distally the fossæ at the anterior bases of the condyles are less pronounced, and the external trochleæ of both condyles are less produced outward. In accordance with this form the proximal phalange is a little narrower proximally than in either foot of the sheep; the shaft is also narrower than that of the sheep's fore foot, while the entire bone is longer than that bone, or the corresponding posterior phalange of the sheep. The ligamentous fossæ are bounded by much more prominent and rugose ridges of insertion. The inferior tendinous insertions are about as in the sheep.

MEASUREMENTS.

					M.
Diameter of middle of shaft of humerus	anteroposterior		.		.018
	transverse				.015
Width of neck of scapula	.	.	.	.	.0175
Length of radius	.	.	.	.	.038
Proximal diameter of radius	vertical (greatest)		.		.013
	transverse	.	.	.	.027
Distal diameter of radius	vertical (of facets)		.		.013
	transverse	.	.	.	.023
Length of anterior cannon bones	.		.	.	.097
Diameter of proximal end of cannon bone	transverse		.		.021
	fore and aft (internally)				.014
Length of external proximal phalange	.		.	.	.037
Diameter " " " proximally	vertical		.		.015
		transverse			.011
" " " " distally	vertical (chord)		.		.011
		transverse (least)		.	.007

Discovered by Dr. Rijgersma with the bones of *Amblyrhiza*, etc.

Homo, Linn.

Associated with the remains described in the preceding pages, Dr. Rijgersma discovered a highly interesting relic of the stone age of the human inhabitants of this portion of the West Indian Islands. I use the term "stone age" in a chronological sense only, since the region in question possesses chiefly coral rock, and little or none that is adapted for conversion into cutting instruments, so that the inhabitants resorted to the use of animal products, as teeth, bones, and shells. The implement found by Dr. Rijgersma is a long-ovate spoon-shaped scraper or chisel, cut by human hands from the lip of the large *Strombus gigas*. The ribs of the external surface and the smooth internal surface are easily distinguished, and the distinct natures of the lamellar and prismatic layers have been evidently well understood by the artificer, who has ground away the latter in order to put a sharp edge on the former at one end. This edge is sharp, and mainly well preserved. The implement has a greater median width, and smoothly ground thick margin; the end of the plate is obtuse and with thick edge, almost entirely composed of the prismatic layer. It has evidently been held in the hand, and been used after the manner of the stone scrapers of the North American Indians.

This chisel was found by Rijgersma under circumstances precisely similar to those attending the discovery of the gigantic rodents. Some portions of each of the species described were imbedded in the breccia, and others occurred loose in a red earth in cavities of the breccia. The chisel has the color and constitution of the latter teeth and bones, and was found with them in this earth. Some of the teeth are even more fresh-looking and less stained than the chisel.

A good figure may be found in the Proceedings of the American Philosophical Society for 1868, Vol. XI. Plate V. fig. 4, which is repeated on Plate I., of the present work.

CONCLUDING REMARKS.

So far as the appearance presented by the specimens described in the preceding pages is an indication, all may have been buried in the cave deposit at one time, and may represent animals which lived together on the Island of Anguilla. But the iron oxide of the deposit evidently imparts so deep a stain to whatever is buried in the cave, that it is impossible to base an estimate of the age of such objects by their color alone. It is well known that no species of *Boridæ* is indigenous to any part of the Neotropical realm, including the West Indian Islands, and that individuals of that genus now in the islands are importations. But the bones of the ruminant of Anguilla are as deeply colored as those of the *Amblyrhiza*, some of them even more so. It is also evident that the implement made from the lip of *Strombus gigas* has been for a long period exposed to the iron oxide, since its stain is deeply fixed, totally replacing the natural color, even to the usual white of the dense enamel of the inner surface seen in fossils of early tertiary age.

Whether the extinct genus *Amblyrhiza* was contemporary with the ruminant and man, is a question which must be left for future investigators. Of one point we may be assured, namely, that the deposit inclosing the *Amblyrhiza* is of late geological age. This may be derived from the close affinity which exists between that genus and existing ones, and also from some species of shells which I found imbedded in the matrix inclosed in the intercondylar fissure of the femur of the *Amblyrhiza latidens*. I sent these to my friend Thomas Bland of New York, whose knowledge of West Indian land-shells is exhaustive, and he informed me that they belong to a species of *Tudora* so nearly allied to the existing species *T. puparformis* as to make it difficult for him to characterize it as distinct. I would not place the age of the *Amblyrhiza* earlier than the Pliocene.

The island of Anguilla, now embracing but thirty square miles, could not readily have supported a fauna of which these huge rodents formed a part. Such large animals have no doubt ranged over a more extended territory. This, and other facts mentioned by Pomel, lend probability to the hypothesis of the latter author, that the submergence of the ranges connecting many of the islands of the Antilles has taken place subsequent to Pliocene times.

EXPLANATION OF PLATES.

PLATE I.

Reptiles, birds, and mammals, all of the natural size, excepting fig. 12.

Fig. 1. Femur of an *Iguana*, posterior view; *a*, exterior; *b*, proximal, and *c*, distal views.

Fig. 2. Humerus of a species of *Procellaria*, front view; *a*, posterior; *b*, proximal, and *c*, distal view.

Fig. 3. Coracoid bone of a large bird, superior view; *a*, interior view.

Figs 4–6. Long bones of a rodent.

Fig. 4. Left humerus without distal extremity, external view; *a*, proximal view.

Fig. 5. Distal portion of right humerus, anterior view; *a*, posterior view.

Fig. 6. Left femur, front view; *a*, posterior view.

Fig. 7. Proximal portion of a rib which accompanied the jaws of *Amblyrhiza latidens*, posterior view; *a*, proximal end.

Fig. 8. Proximal phalange of rudimental first digit of *Amblyrhiza inundata*, superior view; *a*, inferior, *b*, proximal, and *c*, distal views of the same.

Fig. 9. Unknown bone found with those of *Amblyrhiza*; *a* and *b*, extremital views.

Fig. 10. Symphysis mandibuli of *Amblyrhiza* from above.

Fig. 11. Maxillary, molar, and palatine bones of an *Artiodactyle*, supporting the temporary dentition, from the right side; *a*, from below.

Fig. 12. Shell chisel of human manufacture, three-fourths natural size, viewed on the external surface of the shell; *a*, edge view.

PLATE II.

Figs. 1–3. *Amblyrhiza quadraus*, natural size.

Fig. 1. Portion of the left mandibular ramus from above; *a*, the same, the external side; *b*, the inner side, the incisive alveolus broken open.

Fig. 2. Right ramus, probably of the same animal, from above; *b*, the inner side.

Fig. 3. First superior molar from below; *a*, outer side; *b*, inner side.

Figs. 4 and 5. Left and right mandibular rami of *Amblyrhiza latidens*, natural size; lettering as above.

PLATE III.

Mandibular and premaxillary bones of *Amblyrhiza latidens* or *A. quadraus*, natural size.

Fig. 1. Premaxillary bone with portions of both incisor teeth, right side; *a*, superior view.

Fig. 2. Inferior view of premaxillary bone of another individual.

Fig. 3. Right mandibular ramus and symphysis; *a*, from above.

Fig. 4. Symphyseal region with incisor teeth from below; *a*, from above.

(29)

PLATE IV.

Bones accompanying the cranial fragments of *Amblyrhiza quadrans* and *A. latidens*, one-half natural size.

Fig. 1.   Atlas from before; *a*, from behind; *b*, from below; *c*, from above.

Fig. 2.   Axis from left side; *a*, from behind.

Fig 3.   Lumbar vertebral centrum from side; *b*, from front; *c*, from back.

Fig. 4.   Lumbar vertebral centrum; *b*, from front; *c*, from behind.

Fig. 5.   Three vertebræ of sacrum, profile; *a*, from below; *b*, from front; *c* posterior extremity.

Fig. 6.   Caudal vertebra from side; *a*, front view.

Fig. 7.   Proximal end of humerus from behind; *a*, external view; *b*, proximal view.

Fig. 8.   Condyles of humerus broken from the shaft from behind; *a*, proximal view; *b*, distal view.

Fig 9.   Femur without head, external view; *a*, internal view; *c*, section of shaft at fracture.

Fig. 10.   Proximal portion of another femur, with the great trochanter distorted by pressure, posterior view; *a*, internal view; *c*, broken extremity of shaft; *b*, proximal end of shaft.

Fig. 11.   Distal extremity of a femur, right side; *a*, anterior, and *b*, posterior views.

Fig. 12.   Patellar surface of another femur.

PLATE V.

Bones of *Amblyrhiza* and ruminant, one-half natural size.

Fig. 1.   Anterior part of sacrum of *Amblyrhiza ?latidens*, which accompanied the bones of *Amblyrhiza inundata*, viewed from above; *a*, from below; *b*, anterior extremity.

Figs. 2–6.   Bones which accompanied the jaws and teeth of *Amblyrhiza quadrans* and *A. latidens*.

Fig. 2.   Distal portion of right humerus from front; *a*, posterior view.

Fig. 3.   Right acetabulum, external view.

Fig. 4.   Proximal portion of left tibia, interior view; *a*, posterior, *b*, exterior, and *c*, proximal views.

Fig. 5.   Distal portion of left tibia with astragalus in place, anterior view; *a*, posterior view; *b*, distal view of the same tibia alone.

Fig. 6.   Left astragalus from above; *a*, from below; *b*, from the inner side; *c*, distal view; *d*, posterior view.

Fig. 7.   Fourth metatarsal bone, which accompanied the sacrum represented in fig. 1, from the front; *a*, from behind.

Figs. 8–10.   Bones of an adult ruminant described on page 24.

Fig. 8.   Right radius, with parts of the ulna adherent, anterior view; *a*, posterior view.

Fig. 9.   Metacarpal cannon bone, anterior view; *a*, posterior; *b*, proximal; *c*, distal views.

Fig. 10.   Left proximal phalange of the same, from above; *a*, from below; *b*, proximal; *c*, distal views.

PLATE I

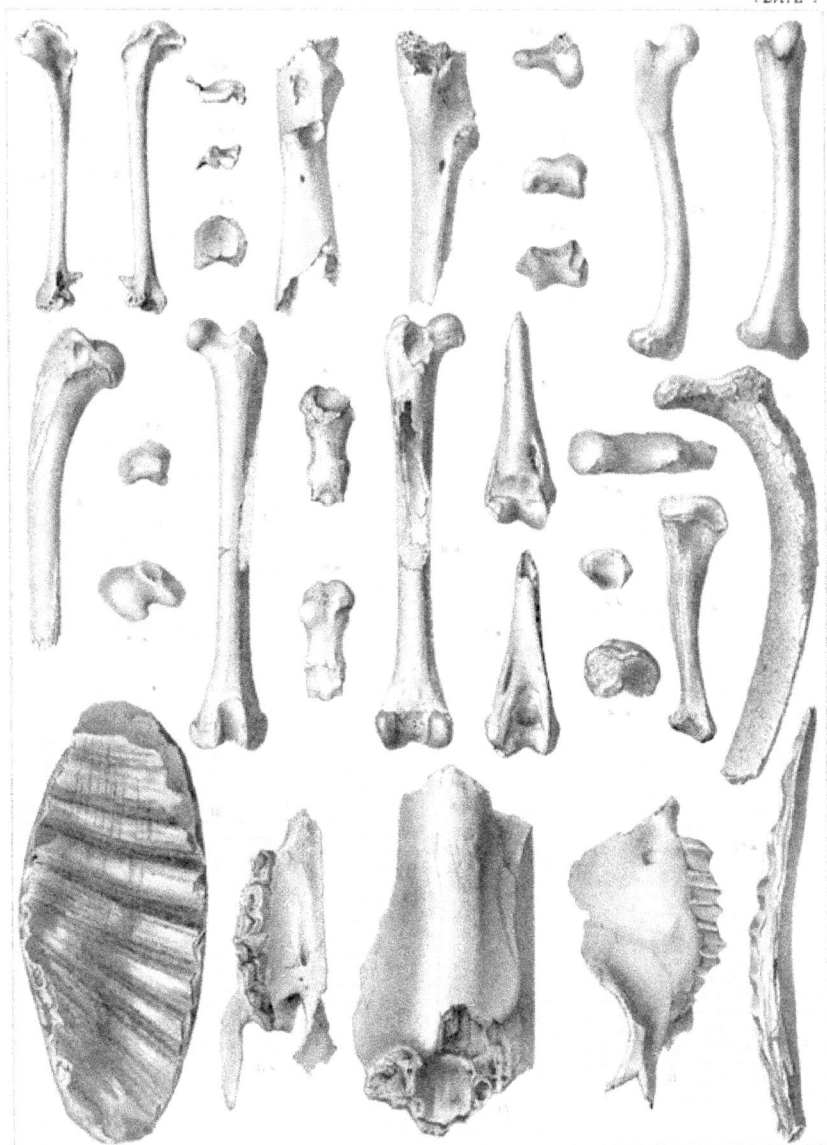

 Sinclair & Son Lith. Phila.

FOSSILS ETC FROM THE ANGUILLA CAVE

PLATE II

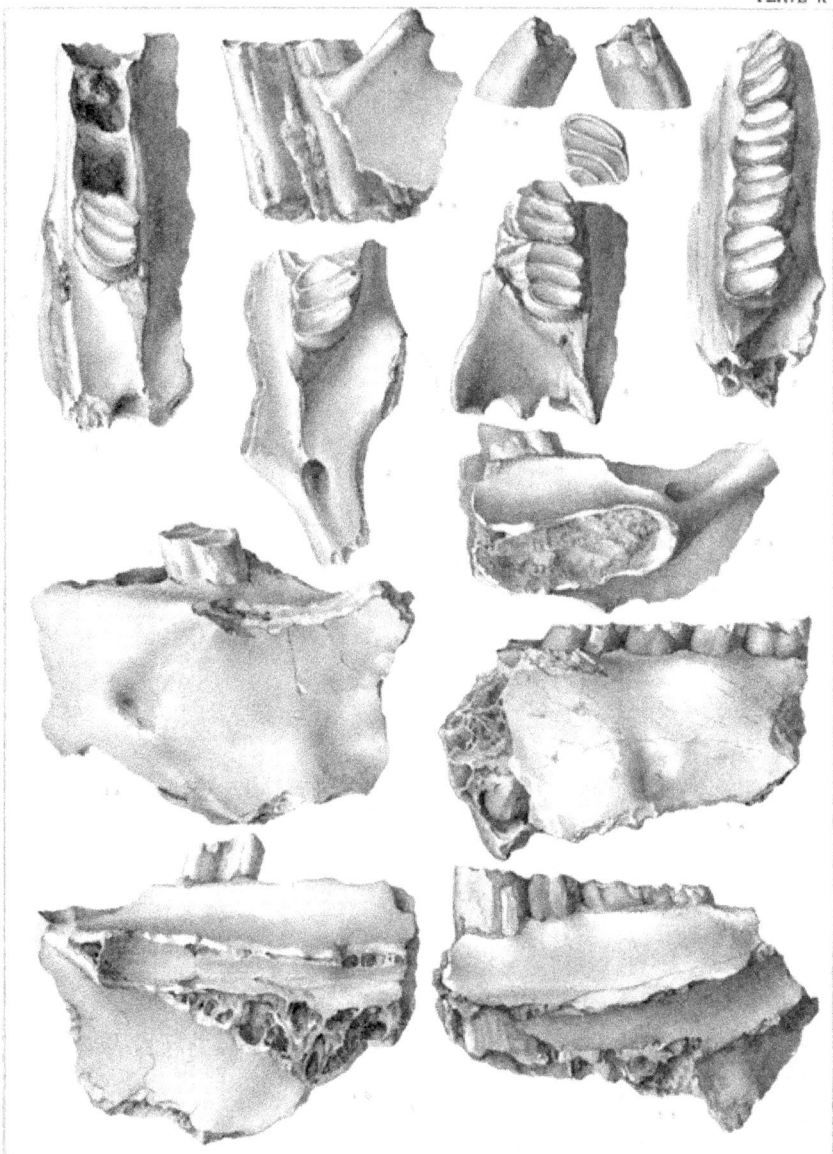

T. Sinclair & Son, Lith. Phila.

1-3 AMBLYRHIZA QUADRANS 4-5 A. LATIDENS. H.

PLATE III

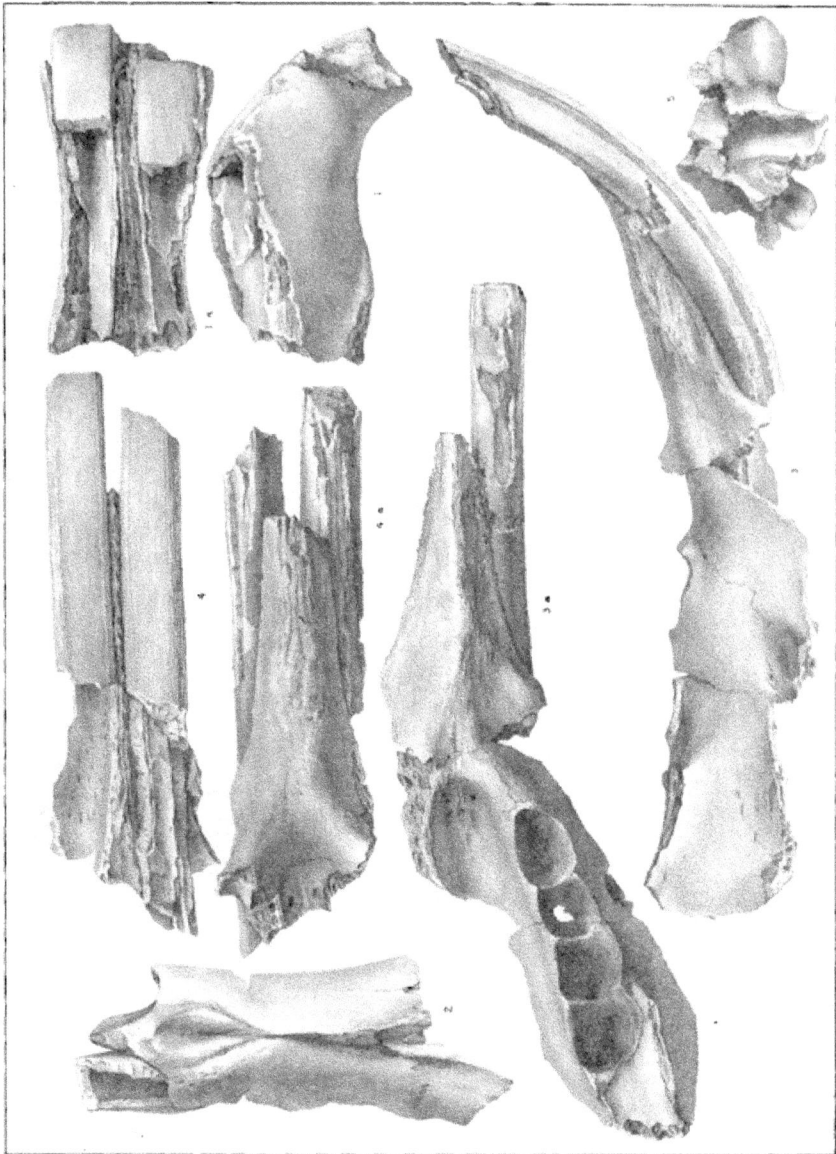

PLATE IV

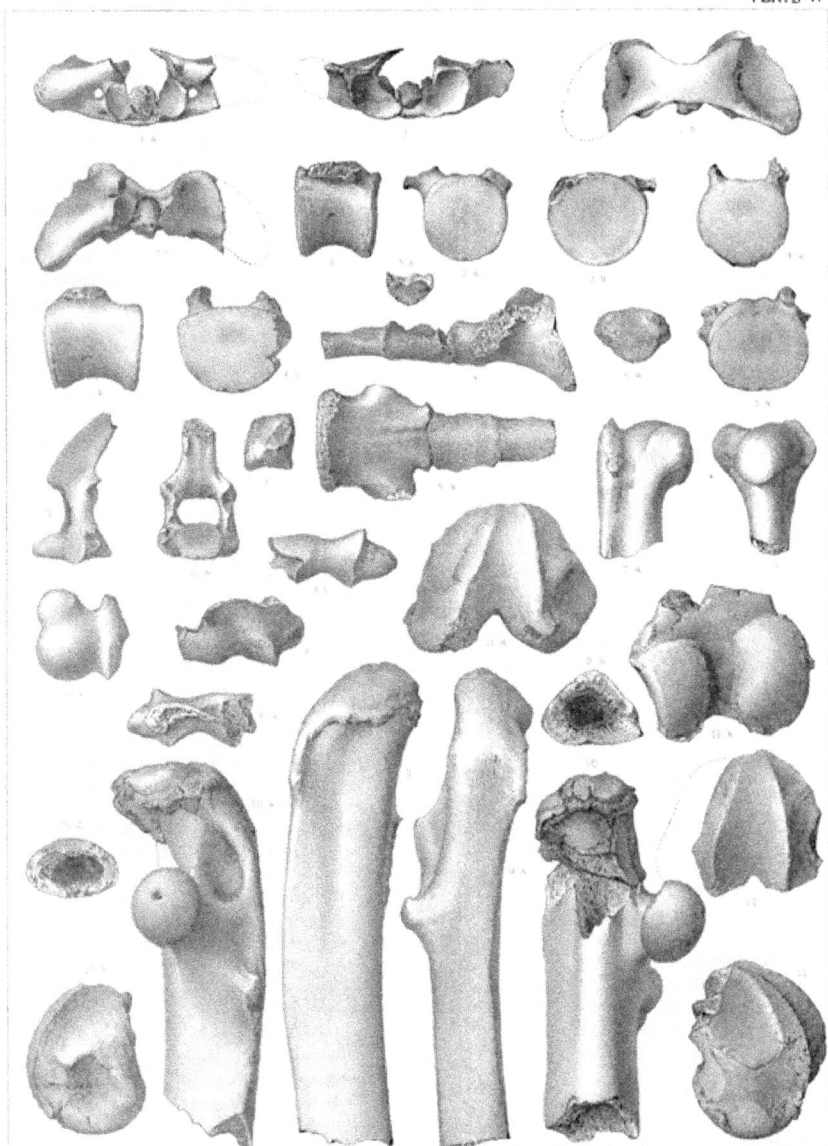

AMBLYRHIZA ½

PLATE V

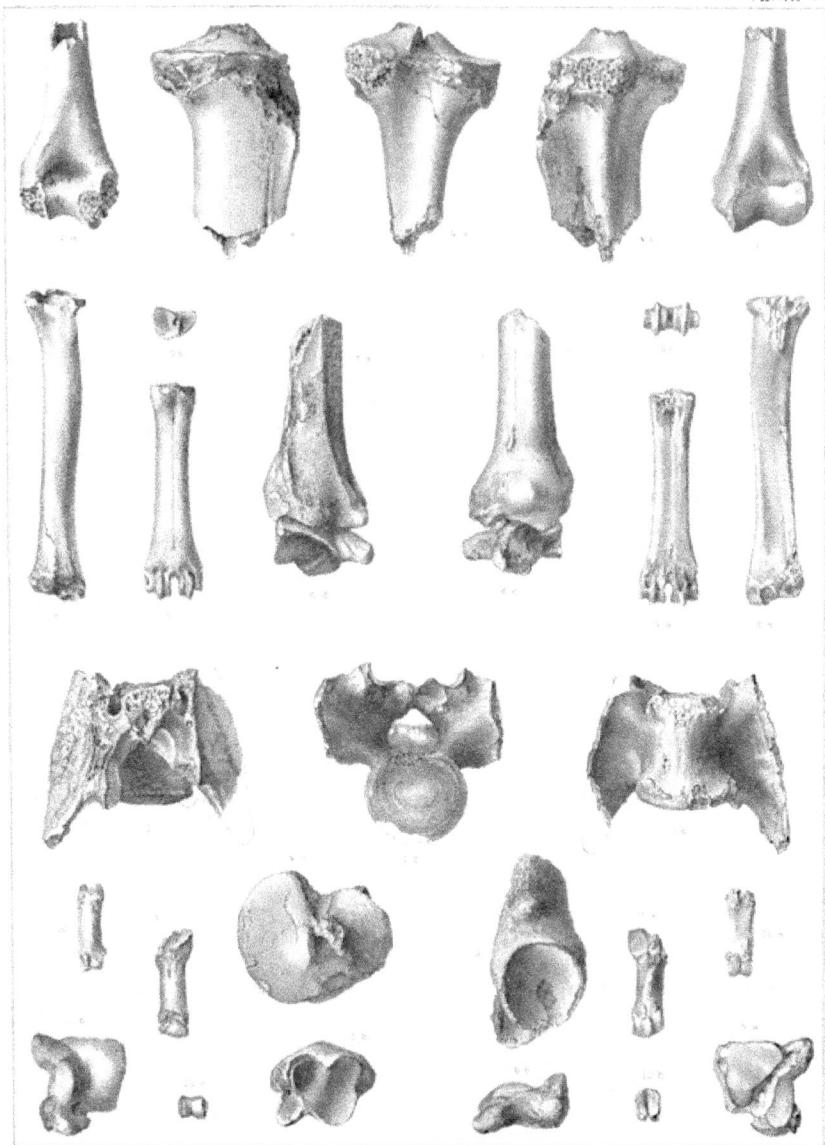

1-7 AMBLYRHIZA 8-10 ARTIODACTYLE ¼

www.ingramcontent.com/pod-product-compliance
Lightning Source LLC
Chambersburg PA
CBHW021548270326
41930CB00008B/1411